QuickStudy.

ELECTRICITY

ELECTRIC FIELDS

An electric field, $\underline{E}$, is estimated as the force, $\underline{F}$, acting on a positive test charge, over the magnitude of the test charge, **q**:

$$\underline{E} = \frac{\underline{F}}{q}.$$

The force, $\underline{F}$, on a negative charge has an orientation opposite to that of the electric field.

- **Coulomb's Law**. The force, $\underline{F}$, exerted between two charged particles, q_1 and q_2, placed at a distance, $\underline{r}$, is:

$$\underline{F} = k\frac{|q_1 q_2|}{r^2}\underline{r}$$

where $\underline{r}$ is the unit vector in the direction of r, and **k** is a proportionality constant whose numerical value is $8.988 \times 10^9\ Nm^2/C^2$. The constant **k** is often written as:

$$k = \frac{1}{4\pi\varepsilon_0}$$

where $\varepsilon_0 = 8.854 \times 10^{-12}\ C^2/Nm^2$.

The force on a particle in an electric field is given by: $\underline{F} = q\underline{E}$.

The magnitude of electric field, **E**, at a field point located at distance, **r**, is:

$$E = \frac{1}{4\pi\varepsilon_0}\frac{|q|}{r^2}.$$

- **Electric Flux**, **FE**, for nonuniform electric fields through an oriented surface, $d\underline{A}$, is defined as $\Phi_E = \iint \underline{E} d\underline{A}$.
- **Gauss Law** states that the electric flux, Φ, equals:

$$\Phi = \oint \underline{E} d\underline{A} = \frac{Q_e}{\varepsilon_0}$$

where Q_e is the total charge under the enclosed surface.

ELECTRIC POTENTIAL ENERGY

The force that acts on a charge moving in a static electric field is a conservative force, and there is an associated potential energy, **U**, given as:

$$U = \frac{1}{4\pi\varepsilon_0}\frac{qq'}{r}$$

where **q'** is the test charge at a distance **r** from the charge, **q**. To find the potential energy, **V**, at a specific point due to any collection of point charges:

$$V = \frac{U}{q'} = \frac{1}{4\pi\varepsilon_0}\sum_i \frac{q_i}{r_i}$$

- **Potential Gradient**. The potential energy at a point is related to the electric field by the equation:

$$\underline{E} = \nabla V = -\left(\frac{\partial V}{\partial x}\underline{i} + \frac{\partial V}{\partial y}\underline{j} + \frac{\partial V}{\partial z}\underline{k}\right).$$

Customer Hotline # 1.800.230.9522
We welcome your feedback so we can maintain and exceed your expectations.

RESISTANCE

- **Ohm's Law**. The current, **I**, is linearly proportional to the potential, **V**, **V = IR** where **R** is the resistance.
- **Resistivity**, ρ, of a material is defined as: $\rho = \frac{E}{J}$ where **E** is the electric field and **J** is the current density. The resistance, **R**, is related to the resistivity, ρ, by the relation:

$$R = \frac{\rho L}{A}$$

where **L** is the length and **A** is the cross sectional area of the conductor.

- **Resistors in Series**. The total resistance, R_T, of **N** resistors connected in series is given as:

Resistors in Series

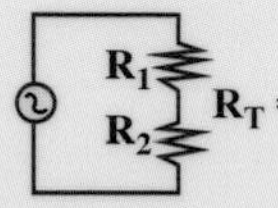

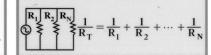

$$R_T = R_1 + R_2 + \ldots + R_N$$

- **Resistors in Parallel**. The total resistance of **N** resistors connected in parallel is given as:

Resistors in Parallel

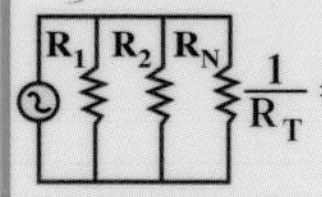

$$\frac{1}{R_T} = \frac{1}{R_1} + \frac{1}{R_2} + \cdots + \frac{1}{R_N}$$

CAPACITANCE

- **Capacitor**: Any two *conductors* separated by an *insulator*. The *capacitance*, **C**, of a system is defined as:

$$C = \frac{Q}{V_{12}}$$

where **Q** is the of charge on either conductor and V_{12} is the potential difference.

The electric field magnitude, **E**, in a *parallel-plate capacitor* (two closely spaced parallel plates in vacuum) is:

$$E = \frac{\sigma}{\varepsilon_0} = \frac{Q}{\varepsilon_0 A}$$

where σ is the surface charge density of each plate, **A** is the surface of each plate, and **Q** is the total charge on each plate. The capacitance of a parallel-plate capacitor in a vacuum is:

$$C = \varepsilon_0\frac{A}{d}$$

where **d** is the distance between the two plates.

- **Capacitors in Series**. The total capacitance, C_T, of **N** number of capacitors connected in series is given as:

Capacitors in Series

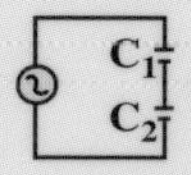

$$\frac{1}{C_T} = \frac{1}{C_1} + \frac{1}{C_2} + \cdots + \frac{1}{C_N}$$

- **Capacitors in Parallel**. T... capacitance of **N** number of c... connected in parallel is given as:

Capacitors in Parallel

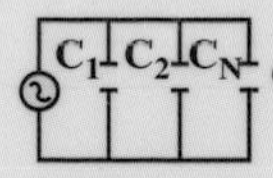

$$C_T = C_1 + C_2 + \ldots + C_N$$

KIRCHHOFF'S RULES

- **Branch Point**. A point where three or more conductors are connected.
- **Loop**. Any closed conducting path.
- **Loop Rule**. The algebraic sum of the potential differences in any loop (including those associated with emfs and those of resistive elements) must equal zero.
- **Point Rule**. The algebraic sum of the currents toward any branch point is zero:

$$\sum_i I_i = 0.$$

MODERN PHYSICS

- **Theory of Relativity**

1. **Lorentz transformation equation** describes an event occurring at a system **S(x,y,z,t)** as experienced by an observer on another system **S'(x',y',z',t')** moving relative to **S** with a constant velocity **u** along the x-axis. **z' = z, y' = y,**

$$x' = \frac{x - ut}{\sqrt{1 - \frac{u^2}{c^2}}},\quad t' = \frac{t - \frac{ux}{c^2}}{\sqrt{1 - \frac{u^2}{c^2}}}$$

where **c** is the speed of light ($u \le c$).

2. **Relativistic Kinetic Energy**

$$K = \frac{mc^2}{\sqrt{1 - \frac{u^2}{c^2}}} - mc^2$$

3. **Total Energy**: $E = K + mc^2$
4. **Decay of Radioactive Nuclei** $N = N_0 \exp(-\lambda t)$ where N_0 the number of nuclei at time zero, and λ is the decay constant.
5. **Quantum of Energy** Electromagnetic wave energy is transported in definite bundles called photons. The energy of a photon is $E = h \times f$ where **f** is its frequency and **h** is Planck's constant and $h = 6.6260 \times 10^{-34} J \times s$.

...ND ...TICS

- **Electromagnetic Forces**. A particle with charge, **q**, moving with a velocity, **v**, through a magnetic field, $\underline{B}$, experiences the **Lorenz force**, $\underline{F}$, $\underline{F} = q\underline{v} \times \underline{B}$. The force, $\underline{F}$, acting on a conductor of length, $\underline{L}$, which carries current, **I**, in a uniform magnetic field, $\underline{B}$, is $\underline{F} = I\underline{L} \times \underline{B}$.

1. **Electromagnetic Moment**. The magnetic moment, $\underline{m}$, of a current loop of directional area, $\underline{A}$, is given as $\underline{m} = I\underline{A}$.
2. **Electromagnetic Torque**. The current loop experiences a torque, $\underline{T}$, defined as $\underline{T} = \underline{m} \times \underline{B}$.
3. **Potential Energy of a Loop** is given as $\underline{U} = -\underline{m} \times \underline{B}$.

- **Magnetic Field**. A particle with charge, **q**, moving with a velocity, $\underline{v}$, produces a magnetic field, $\underline{B}$:

$$\underline{B} = \frac{\mu_0}{4\pi}\frac{q\underline{v} \times \underline{r}}{r^2}$$

where **r** is the distance between the source and the field location, and $\underline{r}$ is the unit vector in the direction of **r**. The constant:

$$\mu_0 = 4\pi \times 10^{-7}\ Wb/(A \times m)$$

1. **Biot-Savart's Law**. A conductor of length, $d\underline{L}$, carrying a current, **I**, produces a magnetic field, $d\underline{B}$:

$$d\underline{B} = \frac{\mu_0}{4\pi}\frac{Id\underline{L} \times \underline{r}}{r^2}.$$

The magnetic field produced by a long, straight wire is given as: $B = \frac{\mu_0}{4\pi}\frac{I}{r}$

- **Maxwell's Equations**. Four equations, that quantify the relationships between electric and magnetic fields.

1. **Gauss Law of Electrostatics**.

$$\oint \underline{E} \times d\underline{A} = \frac{Q}{\varepsilon}$$

2. **Absence of Magnetic Charge**.

$$\oint \underline{B} \times d\underline{A} = 0$$

3. **Ampere's Law**.

$$\oint \underline{B} \times d\underline{L} = \mu_0\left(I_c + \varepsilon_0\frac{d\psi}{dt}\right) = \mu_0(I_0 + I_D)$$

where I_c is the conduction current, I_D is the displacement current, and Ψ is the electric flux.

4. **Faraday's Law**. Where Φ is the magnetic flux.

$$\oint \underline{E} \times d\underline{L} = \frac{d\Phi}{dt}$$

FUNDAMENTAL CONSTANTS AND TERRESTRIAL DATA

QUANTITY	CGS - cent/gram/sec	MKS - meter/kilo/sec	FPS - foot/pound/sec
Gravity (g)	$981 cm/s^2$	$9.81 m/s^2$	$32 ft/s^2$
Speed of Light (c)	3×10^{10} cm/s	3×10^8 m/s	1.8641×10^5 mi/sec
Gravitational Constant (G)	6.67×10^8 Dyne $cm^2\ g^2$	6.67×10^{-11} N $m^2\ Kg^{-2}$	
Gas Constant (R)	8.31451 J $(K \bullet mole)^{-1}$		
Avogadro's Number N_0	6.022×10^{23} molecules/mole	same	same
Boltzman's Constant (k)	1.38×10^{-16} erg K^{-1}	1.38066×10^{-23} J/K	
Electron Volt		1.60219×10^{-19}J	
Electron Charge (c)	4.80×10^{-10} esu	1.60219×10^{-19} coulomb	
Permeability Constant (μ_0)	1.26×10^{-8} henry cm	1.26×10^{-6} henry m	
Permittivity Constant (e_0)	8.85×10^{-14} farad cm	8.85×10^{-12} farad m	
Radius of Earth		6.374×10^6 meters	3959 miles 2.09×10^7 feet
Mass of Earth		5.974×10^{24} kg	
Standard Atmospheric Pressure		1.013×10^5 Pa	14.70lb. $in^2 \Leftrightarrow$ 760mmHg
Speed of Sound Air (20°c)		344 m/s	742.5 miles hr 1089 ft/s

ISBN 157222500-9
9 781572 225008 50495
March 2004 U.S.$4.95 CAN.$7.50
PRINTED WITH SOY INK
6 54614 20500 1

visit us at quickstudy.com

BarCharts, Inc.® WORLD'S #1 ACADEMIC OUTLINE

Quick Study ACADEMIC

PHYSICS

optics vectors mechanics momentum

VECTORS AND COORDINATE SYSTEMS

DEFINITIONS

- **Scalar and Vector Quantities**. Physical quantities such as temperature, **T**, distance, **s**, density, ρ, work, **W**, etc., that can be fully described by a single number are called *scalars*. Scalars are not associated with any direction. Physical quantities that have both ***magnitude*** and ***direction*** are called ***vectors***, e.g, force, $\underline{F}$, velocity, $\underline{V}$, acceleration, $\underline{a}$, momentum, $\underline{P}$, etc.
- **Coordinate Systems**. A vector, $\underline{V}$, can be described in reference to a coordinate system. Two-dimensional coordinate systems can be **Cartesian** or **polar**. Three-dimensional coordinate systems can be **Cartesian**, **cylindrical** or **spherical**.

TWO-DIMENSIONAL (2-D) COORDINATE SYSTEMS

- **Cartesian Coordinates (x,y)**. A vector, $\underline{V}$, in a 2-D Cartesian coordinate system can be written as:

$$\underline{V} = \underline{V}(V_x, V_y) = V_x\underline{i} + V_y\underline{j}$$

where V_x, V_y are the *vector components*, and $\underline{i}$, $\underline{j}$ are the *unit vectors* along the **x** and **y** axis respectively. The *magnitude* of the vector, $|\underline{V}|$, is:

$$|\underline{V}| = \sqrt{V_x^2 + V_y^2}.$$

- **Polar Coordinates (r,θ)**. A vector, $\underline{V}$, in polar coordinates can be written as: $\underline{V} = \underline{V}(V_r, \theta)$, where:

$$V_r = \sqrt{V_x^2 + V_y^2} \text{ and}$$

$$\theta = \tan^{-1}\left(\frac{V_y}{V_x}\right)$$

- **Relation Between Cartesian and Polar Coordinates**:

$$V_x = |\underline{V}|\cos\theta$$

$$V_y = |\underline{V}|\sin\theta$$

$$r = \sqrt{x^2 + y^2}$$

$$\vartheta = \tan^{-1}\left(\frac{y}{x}\right)$$

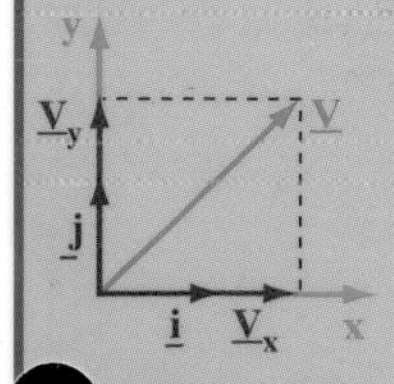

2-D Cartesian Coordinate System

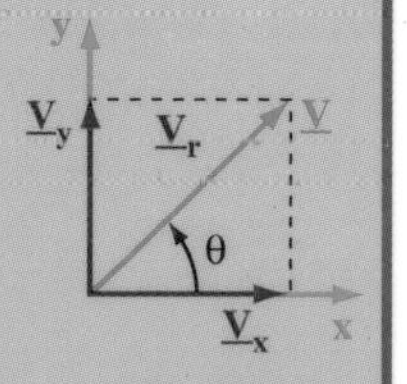

2-D Polar Coordinate System

THREE-DIMENSIONAL (3-D) COORDINATE SYSTEMS

- **Cartesian Coordinates (x,y,z)**. A vector, $\underline{V}$, in a 3-D Cartesian coordinate system can be written as:

$$|\underline{V}| = \sqrt{V_x^2 + V_y^2 + V_z^2}$$

$$\underline{V} = \underline{V}(V_x, V_y, V_z) = V_x\underline{i} + V_y\underline{j} + V_z\underline{k}$$

3-D Cartesian Coordinates

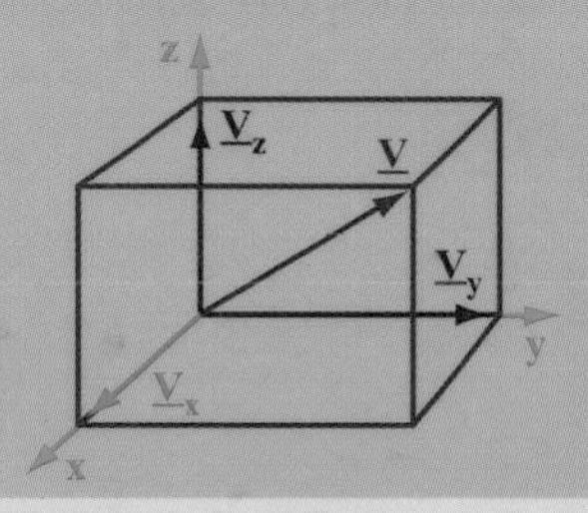

- **Cylindrical Coordinates (r,θ,z)**. A vector, $\underline{V}$, in a cylindrical coordinate system can be written as:

$\underline{V} = \underline{V}(V_r, \theta, z)$, where $V_r = \sqrt{V_x^2 + V_y^2}$ and $z = z$.

Cylindrical Coordinates

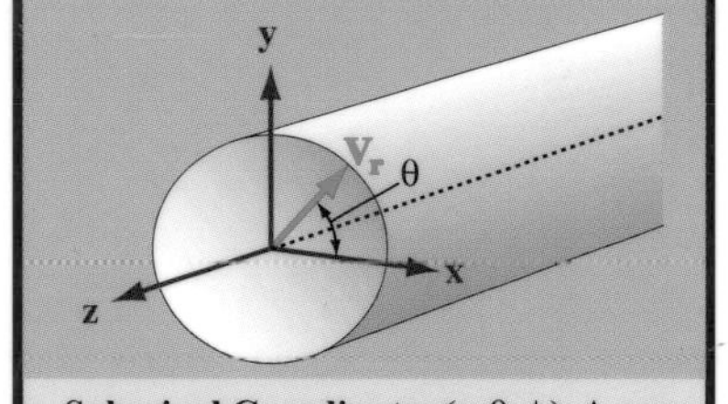

- **Spherical Coordinates (r,θ,ϕ)**. A vector, $\underline{V}$, in a spherical coordinate system is written as: $\underline{V} = \underline{V}(V_r, \theta, \phi)$, where:

$$V_r = \sqrt{V_x^2 + V_y^2 + V_z^2},$$

$$\theta = \tan^{-1}\left(\frac{V_y}{V_x}\right),$$

$$\phi = \cos^{-1}\frac{V_z}{\sqrt{V_x^2 + V_y^2 + V_z^2}}.$$

Spherical Coordinates

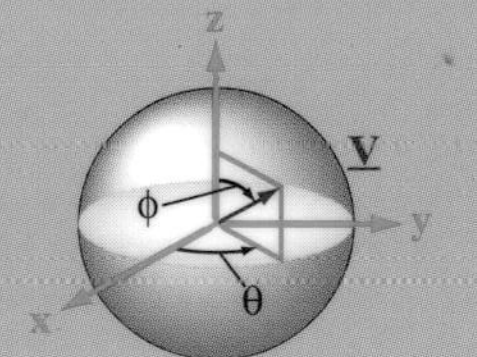

- **Relation Between Cartesian and Spherical Coordinates**

$$V_x = |\underline{V}|\sin\phi\cos\theta$$

$$V_y = |\underline{V}|\sin\phi\sin\theta$$

$$V_z = |\underline{V}|\cos\phi$$

$$V_x = r\sin\phi\cos\theta$$

$$V_y = r\sin\phi\sin\theta$$

$$V_z = r\cos\phi$$

VECTOR ALGEBRA

- **Vector Addition**. The sum of two vectors, $\underline{V}_A$ and $\underline{V}_B$, in a 2-D Cartesian coordinate system is a vector, $\underline{V}_R$, defined as:

$$\underline{V}_R = \underline{V}_A + \underline{V}_B.$$

In component notation, the summation is given as:

$$V_{RX} = V_{AX} + V_{BX},\ V_{RY} = V_{AY} + V_{BY}.$$

Vector Addition

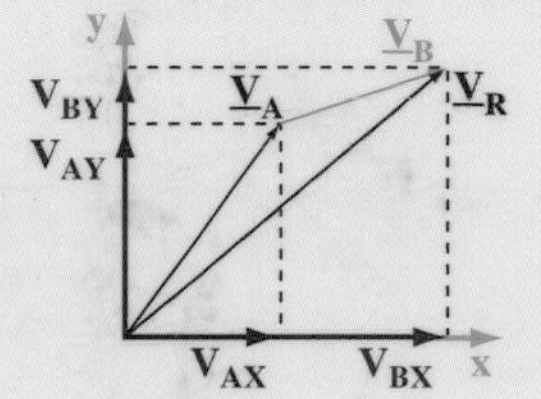

- **Commutative Law of Vector Addition**: $\underline{V}_A + \underline{V}_B = \underline{V}_B + \underline{V}_A$
- **Associative Law of Vector Addition**: $(\underline{V}_A + \underline{V}_B) + \underline{V}_C = \underline{V}_A + (\underline{V}_B + \underline{V}_C)$
- **Distributive Law for Multiplication by a Scalar** (ε): $\varepsilon(\underline{V}_A + \underline{V}_B) = \varepsilon\underline{V}_A + \varepsilon\underline{V}_B$
- **Scalar or Dot Product**: $\underline{V}_A \times \underline{V}_B = \underline{V}_B \times \underline{V}_A = |\underline{V}_A| \times |\underline{V}_B| \cos a$ where α is the angle between the two vectors. If the two vectors are perpendicular to each other, then:

$$|\underline{V}_A| \times |\underline{V}_B| = 0,\ \underline{V}_A \perp \underline{V}_B$$

If the vectors are given in terms of their components, then in a 3-D Cartesian coordinate system:

$$\underline{V}_A \times \underline{V}_B = V_{AX}V_{BX} + V_{AY}V_{BY} + V_{AZ}V_{BZ}$$

since:

$$\underline{i} \times \underline{i} = 1 \quad \underline{j} \times \underline{j} = 1 \quad \underline{k} \times \underline{k} = 1$$

$$\underline{i} \times \underline{j} = 0 \quad \underline{i} \times \underline{k} = 0 \quad \underline{j} \times \underline{k} = 0$$

$$\underline{j} \times \underline{i} = 0 \quad \underline{k} \times \underline{i} = 0 \quad \underline{k} \times \underline{j} = 0$$

Scalar Product

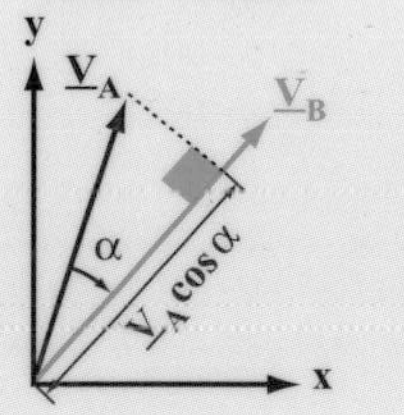

- **Vector or Cross Product** $\underline{V}_A \times \underline{V}_B = |V_A|\,|V_B|\,(\sin a)\,\underline{e}$ where $\underline{e}$ is the unit vector perpendicular to the plane formed by vectors $\underline{V}_A$ and $\underline{V}_B$.
- **Right-hand rule**: The direction of the vector $\underline{e}$ can be found by curling the fingers of the right hand around a hypothetical axis perpendicular to plane $\underline{V}_A - \underline{V}_B$ so the vector $\underline{V}_A$ rotates along the angle α until is aligned with vector $\underline{V}_B$. The thumb then gives the direction of $\underline{e}$.

Right-Handed Rule

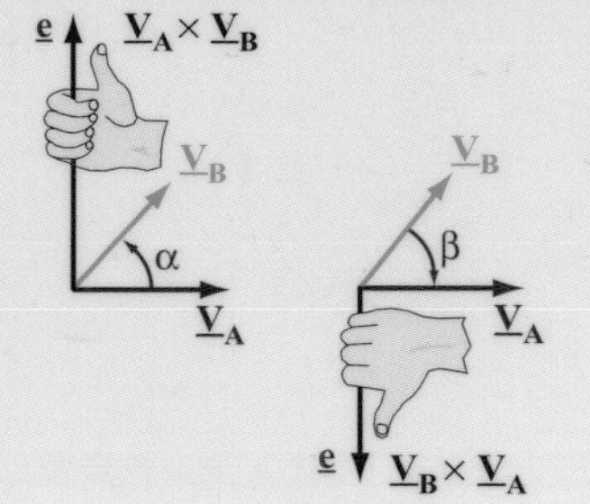

$$\underline{V}_A \times \underline{V}_B = -\underline{V}_B \times \underline{V}_A$$

A Cartesian coordinate system is called a **right-handed system** if $\underline{i} \times \underline{j} = \underline{k}$.

If two vectors are parallel to each other then:

$$\underline{V}_A \times \underline{V}_B = 0,\ \underline{V}_A \parallel \underline{V}_B.$$

If the vectors are given in terms of their components, then in a 3-D Cartesian coordinate system:

$$V_A + V_B = \begin{vmatrix} \underline{i} & \underline{j} & \underline{k} \\ V_{AX} & V_{AY} & V_{AZ} \\ V_{BX} & V_{BY} & V_{BZ} \end{vmatrix}$$

Vector Product

Area $= V_A V_B \sin\alpha$

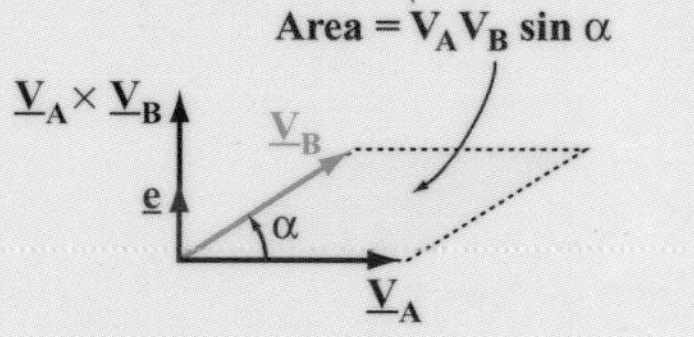

- **Triple Scalar Product** The magnitude of the triple scalar product is equal to the volume of the parallelepiped formed by the three vectors $\underline{V}_A, \underline{V}_B, \underline{V}_C : \underline{V}_A \times (\underline{V}_B \times \underline{V}_C)$.

Triple Scalar Product

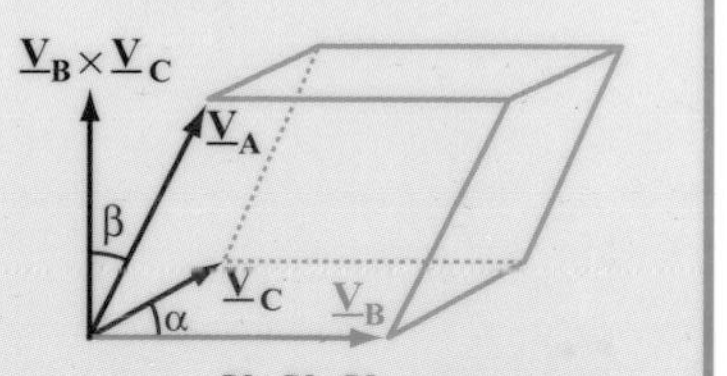

Volume $= V_A V_B V_C \sin\alpha\cos\beta$

- **Differentiation Formulas of Vectors**

$$\frac{d}{dt}\left[\underline{u}(t) + \underline{v}(t)\right] = \frac{du}{dt} + \frac{dv}{dt}$$

$$\frac{d}{dt}\left[c\underline{u}(t)\right] = c\frac{du}{dt}$$

$$\frac{d}{dt}\left[f(t)\underline{u}(t)\right] = \frac{df}{dt}\underline{u} + f\frac{du}{dt}$$

$$\frac{d}{dt}\left[\underline{u}(t) \cdot \underline{v}(t)\right] = \frac{du}{dt} \cdot \underline{v}(t) + \underline{u}(t) \cdot \frac{dv}{dt}$$

- **Integration of a Vector**

$$\int_a^b \underline{r}(t)\,dt = [\underline{R}(t)]_a^b = \underline{R}(b) - \underline{R}(a)$$

NOTE TO STUDENT: This ***QuickStudy***® guide is an annotated outline of the major topics taught in introductory Physics courses. Due to its condensed format use it as a guide, but not as a replacement for assigned class work. © 2002, 2004 BarCharts, Inc. Boca Raton, FL.

MECHANICS

FLUID MECHANICS

- **Definitions**

Fluid: Any substance that under ordinary pressure and temperature conditions cannot sustain shearing stresses without deforming continuously, as long as the stresses are applied.

Liquid: The fluid that can maintain a free-surface boundary.

Gas: The fluid that tends to fill the container in which it is placed.

- **Viscosity: Newtonian Fluids.** For Newtonian fluids the shear stress, τ, and the **rate of shear deformation**, du/dz, are linearly proportional:

$$\tau = \mu \frac{du}{dz}$$

where du/dz is the velocity gradient, **z** is the direction perpendicular to the flow, and μ is the **dynamic** or the **absolute viscosity** of the fluid. All fluids for which $\mu \neq 0$ are called **real fluids**. The fictitious class of fluids for which μ is assumed as zero are called **inviscid**.

- **Hydrostatic Pressure.** The pressure, *p*, of water at rest acting on a point located at depth, **z**, is given as, $p = p_o + \rho_w\, gz = p_o + \gamma_w z$, where p_o is the atmospheric (reference) pressure, ρ_w is the density of water and γw is the specific weight of water.

- **Archimedes Principle.** Every body submerged into a fluid is subject to a *buoyancy force*, **B**, equal to the weight of the fluid displaced by the body, $\mathbf{B} = \gamma_F V$, where γ_F is the specific weight of the fluid.

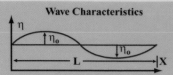

Streamlines

- **Streamline** is a fictitious line in fluid flow such that the velocity of each particle is tangent to that line.

- **Streakline** is the line formed at some instant by a series of fluid particles that have passed through a common point during some previous time interval.

- **Pathline** is the line followed by a fluid particle during a certain time interval.

- **Laminar Flow** is the flow where the streamlines are smooth and orderly and the shear stresses are defined by the equation for Newtonian fluids.

- **Turbulent Flow** is the flow where the streamlines are crossing each other constantly and in a random way.

- **Reynold's Number, Re**, is defined as the ratio between inertial and viscous forces and is the criterion that separates laminar from turbulent flow:

$$Re = \frac{\rho Du}{\mu} = \frac{Du}{\nu}$$

where **D** is a characteristic length (e.g. pipe diameter), **u** is a characteristic velocity, and **v** is the *kinematic viscosity*. Under normal conditions the flow remains laminar for $Re \le 2,000$.

- **Continuity Equation.** The volume flux, **Q**, of an *incompressible fluid* flowing within a pipe is constant: $Q = u_1A_1 = u_2A_2 = \ldots = \text{constant}$ where u_i is the mean flow velocity at a cross sectional area A_i.

- **Bernoulli's Equation.** Neglecting friction losses, the total energy of a flowing fluid is constant:

$$p_1 + \rho gz_1 + \tfrac{1}{2}\rho u_1^2 =$$
$$p_2 + \rho gz_2 + \tfrac{1}{2}\rho u_2^2 = \ldots = \text{constant}$$

where p_i is the pressure and z_i is the elevation.

Fluid Flow

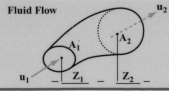

WAVES

- **Wave Characteristics.** *Progressive waves*, are usually characterized by a periodic motion given as:

$$\eta = \eta_0 \sin\,(\Omega t - kx + \alpha)$$

where η_0 is the **wave amplitude**, Ω is the **wave frequency**, **k** is the **wave number** and α is the **phase lag**. The wave frequency and wave number are defined respectively as:

$$\Omega = \frac{2\pi}{T}, \quad k = \frac{2\pi}{L}$$

where **T** is the **wave period** and **L** is the **wave length**.

Wave Characteristics

[diagram with η, η_0, L, X]

- **Phase Velocity.** The **phase velocity** of a travelling wave is defined as:

$$c = \frac{L}{T} = \frac{\Omega}{k}$$

HEAT

Heat can be transferred by means of three different mechanisms, i.e., **conduction, convection** and **radiation**.

- **Conduction.** The thermal energy transfer through a material or between two bodies in contact. Transfer by conduction is due to molecular motion.

1. **Heat Current.** *Heat current*, **H**, is the rate of heat flow, **dQ**, transferred in a certain time period, **dt**: $H = \dfrac{dQ}{dt}$.

The heat current relates the heat transferred through a material of length, **L**, cross sectional area, **A**, and *thermal conductivity*, **k**, as: $H = kA\dfrac{T_H - T_C}{L}$

where $(T_H - T_C)/L$ is the *temperature gradient* (H: hot, C: cold).

2. **Thermal Resistance.** Describes insulating properties of various materials. The thermal resistance, **R**, of a plate or slab of material with thickness, **L**, is defined as: $R = \dfrac{L}{k}$. The heat current is inversely proportional to thermal resistance.

- **Convection.** The transfer of heat by the laminar or turbulent motion of a liquid or gas from one region of space to another.

- **Radiation.** The energy transfer by electromagnetic waves such as infrared, ultraviolet, visible light, etc. Heat current due to radiation is estimated as: $H = A e \sigma T^4$, where A is the surface area, **e** is the **emissivity** of the surface, σ is the **Stefan-Boltzmann constant** and **T** is the *absolute temperature*. The emissivity is larger for dark surfaces than for light. In general, the rate at which a body emits radiation is determined by its own temperature, but the rate at which a body absorbs radiation depends upon the temperature of its surroundings.

1. **Heat Capacity.** The quantity of heat, **Q**, required to increase the temperature of a mass, **m**, of a substance by a small amount, ΔT, is: $Q = mc\Delta T$ where **c** is *specific heat capacity* of the substance.

2. **Phase Changes.** L_f, heat of fusion, defines the amount of heat, **Q**, required per unit mass of a certain substance for a phase change to occur:

$$L_f = \frac{Q}{m}$$ (e.g. from liquid to gas).

THERMODYNAMICS

A **Thermodynamic System** interacts and exchanges energy with its environment in at least two ways, one of which is **heat transfer**. A system in equilibrium cannot change its state without interaction with its environment.

- **Fundamental Laws**

1. **Zeroth Law.** When two systems are in thermal equilibrium with a third system, they are also in thermal equilibrium with each other.

2. **First Law.** When heat, **Q**, is added to a system, some of this added energy remains within the system, increasing its internal energy by an amount, **U**. The remainder of **Q** leaves the system doing work, **W**, against its surroundings: $Q = U + W$. Q is work done on a system while W is work done by a system.

3. **Second Law.** It is impossible to have a cyclic process that converts heat completely into work or a cyclic process that transfers heat from a cooler to a warmer body without requiring work input.

- **Expansion and Compression Work**

$$W_{12} = -\int_{V_1}^{V_2} p\,dV$$

where W_{12} is the work done between states **1** and **2**, **p** is the pressure expressed in absolute units, and **dV** is the change in the system's volume. Pressure may vary during these processes.

- **Thermodynamic Processes**

1. **Adiabatic Process.** A process in which **no** heat transfer into or out of a system occurs: $U_2 - U_1 = \Delta U = -W$.

2. **Isochoric Process.** A constant volume process: $U_2 - U_1 = \Delta U = Q$.

3. **Isobaric Process.** A constant pressure process: $W = p(V_2 - V_1)$.

4. **Isothermal Process.** A constant temperature process.

- **The Ideal Gases**

1. **Ideal (Perfect) Gas Equation**

$$pV = NkT = nRT$$

where **N** is the number of molecules or particles in one mole of gas, **k** is the Boltzmann's constant, **n** is the number of moles and **R** is the specific gas constant defined as: $kN = R$ (*N is Avogadro's number*).

2. **Internal Energy.** The internal energy, **u**, of an ideal gas is given as:

$$u = u\left(\frac{1}{p}, T\right) = u\left(v_s, T\right).$$

The change in the *internal energy*, Δu, is only a function of temperature: $\Delta u = \int c_v\,dT$, where c_v is the *specific heat at constant volume*.

3. **Enthalpy.** The property of **enthalpy**, **h**, is defined as:

$$h = u + pv_s = u + p\frac{V}{m}$$

where **u** is the internal energy and v_s is the specific volume. The change in enthalpy is only a function of temperature: $\Delta h = \int c_p\,dT$ where c_p is the **specific heat** at constant pressure.

4. **Carnot's Law.** For ideal gases c_p and c_v are constants, therefore: $c_p - c_v = R$, For monatomic gases $\Delta u = c_v \Delta T$ and $\Delta h = c_p \Delta T$

- **Entropy.** The change in **entropy**, ΔS, is a reversible process defined as:

$$\Delta S = \int_1^2 \frac{dQ}{T}$$

OPTICS

Geometric Optics is the branch of optics which represents waves in terms of rays.

- **Reflection:** The bouncing of the path of an incident ray of light falling on a smooth surface. **Specular** reflection is perfect reflection at a specific angle from a very smooth surface. **Diffuse** reflection is scattered reflection from a rough surface.

- **Refraction:** The change of direction of a ray of light passing from one medium to another. The optical properties of a medium are described by its *index of refraction*, **n**, given as: $n = \dfrac{c_0}{c}$ where c_0 is the speed of light in a vacuum, and **c** is the speed of light within the medium. The index of refraction for vacuum is **n=1**. The incident, reflected and refracted rays and the normal to the interface are all coplanar.

- **Law of Reflection.** The angle of incidence, θ_i, and the angle of reflection, θ_{rf}, are equal for all wavelengths and for any pair of media: $\theta_i = \theta_{rf}$.

- **Snell's Law** (Law of Refraction). The ratio of the sines of the angles θ_i and θ_r, measured from the normal to the interface is equal to the inverse ratio of the two indices of refraction:

$$\frac{\sin \theta_i}{\sin \theta_r} = \frac{n_r}{n_i}$$

Reflection and Refraction

[diagram: Incident Ray at angle θ_i, Reflected Ray at $\theta_r f$, Refracted Ray at θ_r, with n_i n_r]

CONVERSION FACTORS

LENGTH (Small)

	cm	m	in	ft	yd
cm	1	.01	0.3937	3.2808×10^{-2}	1.0936×10^{-2}
m	100	1	39.370	3.2808	1.0936
in	2.54	2.54×10^{-2}	1	8.3333×10^{-2}	2.7777×10^{-2}
ft	30.48	3.048×10^{-1}	12	1	3.3333×10^{-1}
yd	91.44	9.144×10^{-1}	36	3	1

LENGTH (Large)

	km	m	LY	AU
km	1	0.6214	1.0570×10^{-13}	6.6845×10^{-9}
mile	1.6093	1	1.7011×10^{-13}	1.0757×10^{-8}
LY	9.4605×10^{12}	5.8787×10^{12}	1	6.324×10^4
AU	1.49599×10^8	9.2961×10^7	1.5813×10^{-5}	1

MASS (or weight at sea level g =9.81m/s²)

	g	kg	lb	oz
1 g	1.0000	0.001	2.2×10^{-3}	3.526×10^{-2}
1 kg	1000.0000	1.000	2.2040	3.526×10^{-5}
1 lb	453.5900	0.454	1.0000	16.000
1 oz	28.38	2.84×10^{-2}	0.0625	1.000

PARTICLE KINEMATICS

- **One-Dimensional Rectilinear Motion.** Whenever a particle moves along a straight line, the motion is called **rectilinear**. For rectilinear motion, the **displacement**, x; **velocity**, **v**; **acceleration**, **a**; and **time**, **t**; are related by the following equations:

$$v = \frac{dx}{dt}, \quad a = \frac{dv}{dt} = \frac{d^2x}{dt^2} = v\frac{dv}{dx}$$

for constant acceleration:

$$x = x_0 + v_0 t + \tfrac{1}{2}at^2, \quad v = v_o + at$$
$$v^2 = v_o^2 + 2a\,(x - x_o)$$

where x_o and v_o are the displacement and the velocity at time **t = 0**, respectively.

- **Two-Dimensional Rectilinear Motion.** Rectilinear motion with constant acceleration, **a**, in a 2-D Cartesian system can be described in terms of its **x - y** components as:

$$x = x_o + v_{ox}t + \tfrac{1}{2}a_x t^2$$
$$y = y_o + v_{oy}t + \tfrac{1}{2}a_y t^2$$

where: $v_{ox} = |v_o|\cos\theta$ and $v_{oy} = |v_o|\sin\theta$ and θ is the angle formed between the x-axis and the **initial velocity** vector $\underline{V}_o$.

2-D Rectilinear Motion

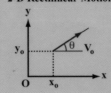

- **Circular Motion.** A particle moving in a circular path at a radius, **R**, from a center point, **O**, will have a *centripetal acceleration*, a_c, directed towards the center of the circle with magnitude:

$$a_c = \frac{u^2}{R} = \omega^2 R = u\omega$$

where **u** is the velocity and ω is the *angular velocity* of the particle. The time, **T**, required for the particle to make one complete revolution is called **period**. Thus the velocity, **u**, can be written as:

$$u = \frac{2\pi R}{T} = \omega R$$

The particle may also have a component of acceleration, a_t, in the direction of its forward motion:

$$a_t = \frac{du}{dt} = \frac{d\omega}{dt}R$$

Circular Motion

FRICTION

Friction, F_f, is a force which inhibits the movement of objects, acting in the opposite direction of the object's motion:

Frictional Force

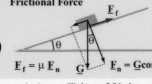

$$F_f = \mu F_n \qquad F_n = G\cos\theta$$

where μ is the coefficient of friction, and F_n is the normal force between the object and the surface in contact. If there are no external forces, for a flat surface $\underline{G} = \underline{F}_n$.

NEWTON'S LAWS OF MOTION

- **First Law.** A body at rest will remain at rest, and a body in motion will remain in motion, unless it is compelled to change its state by forces acting on it.

- **Second Law.** The sum of forces, $\underline{F}$, acting on a body is equal to its mass, **m**, times its acceleration, $\sum \underline{F} = m\underline{a}$.

- **Third Law.** To every action there is an equal and opposite reaction.

- **Type of Forces.** Forces are classified as **body forces** if they act on the entire mass, e.g., gravity, or **surface forces** if they act only upon the surface, e.g. friction.

STRESS & STRAIN

Strain, ε, is defined as the deformation per unit length, where δ is the deformation and **L** is the total length.

Stress is defined as the force per unit area.

Strain

[diagram: L, δ, F_n] $\varepsilon = \dfrac{\delta}{L}$

- **Normal Stress.** Normal stress, σ, acts perpendicular to the surface: where $d\underline{F}_n$ is the elementary normal force acting on the infinitesimal surface area **dA**. If $d\underline{F}_n$ is directed towards the surface then the body is under **compression (negative strain)**. If $d\underline{F}_n$ is directed outwards from the surface, then the body is under **tension (positive strain)**.

Normal Stress

[diagram: F_n, dA] $\sigma = \dfrac{dF_n}{dA}$

- **Shear Stress.** Shear stress, τ, acts tangentially to the surface: where the elementary *tangential force* $d\underline{F}_t$ acting on the infinitesimal surface dA.

Shear Stress

[diagram: dA, F_t] $\tau = \dfrac{dF_t}{dA}$

- **Elasticity** is the property of a material to return completely to its original shape after the loading is removed.

- **Hook's Law.** States that for *elastic bodies* stress is linearly proportional to strain: $\sigma = E\varepsilon$, where **E** is the **modulus of elasticity**. For an elastic spring, stretched or compressed by a force, $\underline{F}$, to a distance, $\Delta\underline{x}$, Hook's Law can be written as: $\underline{F} = \kappa\Delta\underline{x}$ where κ is the **spring constant**.

GRAVITATIONAL FORCE

- A gravitational attraction force, $\underline{F}$, exists between any two bodies of mass m_1 and m_2.

$$\underline{F} = G\frac{m_1 m_2}{R^2}\underline{r}$$

where **R** is the distance between the two masses, $\underline{r}$ is the unit vector in the direction of the force, and **G** is the gravitational constant. The acceleration, **g**, due to earth's gravity, given by the expression:

$$g = G\frac{m_E}{R_E^2}$$

where R_F and m_F are the radius and mass of the earth.

- **Weight.** The weight of a body, **G**, is a body force defined as:
$$\underline{G} = m\underline{g} = \rho V\underline{g} = \rho g V\underline{k} = \gamma V\underline{k}$$
where γ is the **specific gravity**, **g** is the acceleration due to gravity and $\underline{k}$ is the unit vector in the direction of gravity.

WORK & ENERGY

- **Kinetic Energy.** The kinetic energy, **K**, of a particle of mass, **m**, moving with constant velocity, **v**, is defined as:

$$K = \tfrac{1}{2}mv^2$$

The rotational kinetic energy of a body rotating with constant angular velocity, ω and having moment of inertia, **I**, is given as: $K = \tfrac{1}{2}I\omega^2$.

- **Potential Energy.** The potential energy, **U**, of a particle of mass, **m**, positioned at height, **h**, is defined as: $U = mgh$. The potential energy of a spring compressed by a distance, Δx, is:

$$U = \tfrac{1}{2}\kappa\left(\Delta x\right)^2$$

where κ is the spring constant.

If there is no work done, energy is conserved. However, the available energy will probably be partitioned between kinetic and potential energy in a different way at the end of the process than at the beginning. If there is *work*, W_{12}, done, then:

$$K_1 + U_1 = K_2 + U_2 + W_{12}$$

where subscripts (1) and (2) denote the initial and final state of the process respectively.

- **Work.** The work, **W**, done on a particle by a force, $\underline{F}$, is equal to:

$$W = \int_1^2 \underline{F} \times d\underline{x}$$

where $d\underline{x}$ is the distance vector. If the force is constant, and in the same direction as the displaced **S**, the work is: $W = |F|s$.

- **Power.** Power, **P**, is defined as the time rate of doing work $P = \dfrac{dW}{dt}$.

The instantaneous power of a force acting on a particle at velocity, $\underline{v}$, is:

$$P = \underline{F} \times \underline{v}$$

- **Particle Collision.** For an **elastic collision**, the energy of colliding elements remains the same before and after the collision. For an **inelastic collision**, part or all of the energy is lost during the collision due to deformation of the objects and production of heat, sound, etc.

IMPULSE AND MOMENTUM

- **Impulse.** The impulse, $\underline{J}$, of a force, $\underline{F}$, acting over time, Δt, is defined as: $\underline{J} = \underline{F}\Delta t$ where $\underline{F}$ remains constant throughout the time Δt.

- **Momentum.** The momentum, $\underline{P}$, of a body of mass, **m**, moving with velocity, $\underline{u}$, is: $\underline{P} = m\underline{u}$. Momentum is conserved for both elastic and inelastic collisions. The difference between momentum of a body at two instances equals to the impulse:

$$\underline{J} = \underline{P}_2 - \underline{P}_1, \text{ or } \underline{F}\,\Delta t = m\,(\underline{u}_2 - \underline{u}_1)$$

- **Center of Mass.** The coordinates, **X - Y**, for the center of mass of a system of *N* particles are defined as:

$$x = \frac{m_1 x_1 + m_2 x_2 + \cdots + m_n x_n}{m_1 + m_2 \cdots + m_n}$$

$$y = \frac{m_1 y_1 + m_2 y_2 + \cdots + m_n y_n}{m_1 + m_2 \cdots + m_n}$$

where $x_i - y_i$ are the coordinates of the particle m_i.

ROTATIONAL KINEMATICS AND

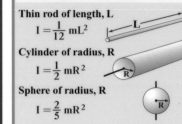

$$\omega = \frac{d\theta}{dt}, \quad \alpha = \frac{d\omega}{dt} = \frac{d^2\theta}{dt^2}$$

The kinematic equations for displacement, θ, and velocity, ω, are similar to those for linear motion:

$$\theta = \theta_0 + \omega_o t + \tfrac{1}{2}\alpha t^2, \quad \omega = \omega_o + \alpha t$$
$$\omega^2 = \omega_o^2 + 2\alpha\,(\theta - \theta_o)$$

The relationships between linear and angular velocity and acceleration are:

$$v = \omega r \text{ and } a = \alpha r$$

where **r** is the distance of the reference point from the axis of rotation.

- **Moments of Inertia.** The moment of inertia of an object of total mass, **m**, about a given axis, x, is defined as: $I_x = \sum r_i^2\, m_i$ where r_i is the distance of the elementary mass m_i from the axis of rotation. In addition, $m = \sum m_i$. The moment of inertia for some very common shape homogeneous objects is as follows.

Thin rod of length, L
$$I = \frac{1}{12}mL^2$$

Cylinder of radius, R
$$I = \frac{1}{2}mR^2$$

Sphere of radius, R
$$I = \frac{2}{5}mR^2$$

- **Torque.** *Torque*, **T**, is the tendency of an object to rotate about a point: $\underline{T} = \underline{r} \times \underline{F}$, where $\underline{r}$ is the position vector and **F** is the force. The sum of all torques about any point of an object in equilibrium is equal to zero. If the object is accelerating with an angular acceleration, α, then: $\sum T = I\underline{\alpha}$.

- **Angular Momentum.** The angular momentum, $\underline{L}$, of an object is defined as the momentum of that object with respect to the point of rotation $\underline{L} = \underline{r} \times m\underline{v}$. The angular momentum is related to the torque by the following equation:

$$\underline{T} = \frac{d\underline{L}}{dt}$$

PERIODIC MOTION

A body is subject to periodic motion if its displacement, x, is proportional to a restoring force. Generally, a periodic motion is described by a sinusoidal function as: $x = x_0 \sin(\Omega t + \theta)$, where x_0 is the amplitude of the oscillation, Ω is the frequency, and θ is the phase lag between the restoring force and the displacement. The frequency is given as: $\Omega = \dfrac{2\pi}{T}$ where **T** is the period of oscillation. For a spring of mass, **m**, and spring constant, κ, the frequency is given by: $\sigma = \sqrt{\dfrac{\kappa}{m}}$.

For a pendulum of length, **L**, the frequency is: $\sigma = \sqrt{\dfrac{g}{L}}$.

DENSITY

The **density**, ρ, of a body is defined as: $\rho = \dfrac{dm}{dV}$. **dm** is the elementary *mass* and **dV** is the elementary volume. For homogeneous systems, the elementary mass and volume can be replaced by the total mass and total volume, respectively.